Mercenary Operations Manual

by Franklin J. Camper

The Combat Bookshelf

Desert Publications

El Dorado, AR 71731

Mercenary Operations Manual

by

Franklin J. Camper

Desert Publications
P.O. Box 1751
El Dorado, AR 71731
501-862-2077

ISBN #:0-87947-241-3

First Printing 1986
Second Printing 1991

Desert Publication is a division of
The DELTA GROUP, Ltd.
Direct all inquires & orders to the above address.

Table of Contents

A few thoughts before you pick up a rifle and go—

* You are not Superman.
* If the enemy is in range, so are you.
* Try to look unimportant. They may be low on ammunition.
* Teamwork is essential. It gives them someone else to shoot at.
* "Friendly" fire isn't.
* When in doubt, empty the magazine.
* The easy way is always mined.
* Close only counts if you're on the receiving end.
* Never draw fire. It irritates everyone around you.
* If you can't remember, you left the Claymore pointed in your direction.

You are on your own. . .

The contents of this manual are intended as information, and the Mercenary Association assumes no responsibility for your use or employment of any of the material herein.

The Association does not make or recommend any suggestion that you break a law or engage in an unlawful activity. In order to know the difference between what is currently legal or illegal, consult with your attorney or an attorney in the country of your operation.

Keep your head down, and good luck.

INTRODUCTION

Those who have tried to get merc jobs have discovered the average mercenary position is only a little harder to find than a small needle in a large haystack. What makes it rough is sometimes they cheat and don't put the needle in there at all.

Hot ads turn out to be gimmicks to get your money. Sure tips lead to dead ends or cold trails. Promises of recruitment fall to pieces under examination. But even with the lies and disappointments, there are those who keep searching, lured by color photographs in the paramilitary magazines of mercs plying their trade in exotic locales . . . and sometimes it seems even those two-dimensional figures with their foreign weapons and suntans are no more real than ghosts who live between the pages of certain magazines.

Well fella, we're here to tell you not only is mercenary work available out there, it's as exciting, dangerous, and demanding as the fiction thrillers make it out to be.

The catch is you have to qualify; and often people get killed permanently dead trying to do it.

THE MERCENARY SOLUTION

The merc exists because there is a ***need.***

Most people simply assume mercenaries are supplements to or replacements for government troops. In the most strict sense of the definition, this is probably fact, but this fact, for being told, obscures the truth.

Mercenary soldiers are the only answer for governments who cannot or will not carry out their own military responsibilities.

There are many missions for which the officers and men of conventional forces are not trained, missions they find repulsive for cultural, moral, or religious reasons, or missions they cannot perform for political reasons.

The merc is often the only means by which a country that needs a job done—but ***by its own self-imposed limitations on its own troops***—cannot do it.

We'll use the United States of America as our example, but the rule applies to every country.

The USA has in place now a military and governmental bureaucracy of incredible size and power. ***Bureaucracy does not find ways to do things. It finds reasons not to do them.***

This means we have spies who are not allowed to spy, soldiers who are not allowed to fight, and adequate punishment within the system for those who try to act on their own initiative.

The military/governmental bureaucracy rattles its sabres by purchasing hardware in the form of more advanced and expensive tanks, missles, aircraft, and thinks this sufficient. Its troops cannot maintain the equipment, its suppliers are corrupt and fail to design usable gear, and its ranks are filled with men and women who are not allowed to act or accomplish.

The Son Tay prison camp raid (no one home when they arrived) and the still classified fiasco at Desert One in Iran, are the two best and most recent examples of commando exercises by bureaucratic committee control.

Enter the Mercenary.

Sometimes we can get another country to do it for us, perhaps one of the more realistic smaller countries who cannot yet afford our system of mismanagement.

Israel and South Africa have served the real interests of the United States well over the last decade in this capacity, raiding, defending and invading.

But even if our friends are unwilling, or their own restrictions get in the way, who is there to shoot the rifle?

Right. The merc. There are two methods the U.S. has of "encouraging" mercenary activities—

1. Ignore them and allow the operation to continue.
2. Assist the program.

By ignoring the operation (usually meaning the op is paid for or controlled by a close ally of the U.S., and serves the interests of both countries), the U.S. government applies its blessing, and lets it continue. This is the most favored bureaucratic way.

To actually assist sends chills up and down the spines of government agencies. It implies ***they are involved,*** and an agency would far rather see something not happen and have no trouble, than to see something tried with the possible threat of failure.

For the U.S. government to help an operation, there must be a dire and immediate disaster on the horizon even the most myopic, deskbound buck-passer recognizes.

Often, government assistance really means government ***control.***

Crippling of the mission or sometimes outright failure is almost assured when governmental control is imposed. The reason is simple. In order to assure no one will be offended, (national boundries, etc.) and that the mission ***will*** succeed, (complications to the plan, presence of U.S. intelligence officers in charge, etc.) the mission is detailed and ridden to death before it is attempted . . . and the burden of being ***helped*** slows everything down, makes plans unchangeable, and turns a simple mission into a disaster.

Official commando units, like the Delta force, organized within the U.S. military are often only salve to a deep wound, something the government can point to and say, "See, we're taking action."

It is not that our special units are not staffed with brave men and not that they are untrained or incapable; they are excellent, but they must follow the dictates of self-serving politicians and timid bureaucrats who have the say if those capable men can go and do something about a bad situation. Under this type of oppression, good units go sour—and its men leave and often become merc later, sworn to ***never*** relive the mistakes they saw while on active duty.

Why mercenaries?

By now the answer should be obvious.

THE TECHNICAL MERCENARY

Taking a brief recon back through history, we see the first mercenaries were men hired in mass, not just by governments and royalty, but by individuals for defense and conquests, contracting out as armies-for-hire.

Back then, you could defend an entire nation with an army of mercs, or pad your own military with a few extra brigades if the royal treasury permitted.

As time passed, and the military arts specialized, large groups of mercenaries went out of fashion. When governments discovered they could impress the farmer and workaday peasant into the national army for little cost, the demand dropped for ready standing troops of mercs.

But specialization brought changes. Highly skilled military men could sell their skills to armies in need, so the merc as infantry soldier began to slowly change to the merc as ***specialist.*** This puts the merc into an individual bargaining position where his predecessors had hired out in groups.

This brings us to the present situation. The free world merc is usually a specialist, acting alone or with small, technical teams. He can be working with a mass of troops of course, but he will be there as an advisor or trainer.

This description would seem to place the mercenary in a position as military-executive for hire, available to improve the quality of his employers army, but there is more to it than that.

Military specialty fields that later pay off in mercenary contracts are;

1. Electronics/missles/radar
2. Aircraft (especially helicopters)
3. Tracked/armored/vehicle repair
4. Logistics/supply
5. Armory/small arms/machine gun repair
6. Indirect fire weapons skill (mortar-howitzer)
7. Special Forces/Ranger combat experience

Note that a very popular field, SF/Ranger training, is at the bottom of the list. Combat personnel, meaning rifle-carrying, bush-living types, are often a low priority on the market. A helicopter mechanic with tool box, will usually get to a foreign combat zone much quicker than a beret-topped lad with a submachine gun.

This is not to say there are not people in there on the ground shooting at each other . . . but the shooting is usually done by cheaper local talent.

MERCENARY TRAINING

Where do you go for training as a merc? Is military experience enough? Are the private schools that teach combat and survival any good?

These are questions often asked by people who want to get into mercenary work, and they are valid questions, the answers sought by serious contenders.

STATESIDE COMBAT AND SURVIVAL SCHOOLS

We have a fair selection of schools in the U.S. that say they teach military combat and field survival skills. Look in the classified section of the paramilitary magazines and you'll find them.

They break down into distinctive groups.

1. Shooting and Firearm schools

2. Bodyguard schools
3. Hand-to-Hand combat schools
4. Field survival and general fieldcraft schools
5. Paramilitary, or Mercenary schools

SHOOTING SCHOOLS

The Shooting Schools are weapons courses designed to teach you pistol, rifle, or shotgun handling. You do a lot of shooting there, as the weapons are the stars of the show. If you want to improve your small arms skills, book yourself into a shooting school.

This sort of school will put you in many firearms usage situations, allowing you to expand your ability by firing under controlled conditions at all types of targets. Technical functioning of the weapons will be taught, and information about modifications, accessories, and ammunition will be expounded. You can learn a lot at a good shooting school.

BODYGUARD SCHOOL

Bodyguard schools are rare, and first-rate ones are even more rare. There is more to bodyguarding than meets the casual eye. It is an ultraspecialized job, and requires knowledge and practical skill in the use of handguns, unarmed combat, alertness, automobiles, travel and hotel security, public relations, and medical lifesaving techniques.

Discipline is the prime asset of the bodyguard. A good school to instruct the ways and means of this type of protection stresses that discipline. While subjects like firearms will be covered in bodyguard school, it is advisable for the aspiring bodyguard to take side courses in related schools like shooting or hand-to-hand combat.

HAND-TO-HAND SCHOOLS

Hand-to-Hand schools are probably the oldest type of schools we have available to us. They began as Karate, Judo, and Ju-Jitsu courses, and evolved to more obscure and exotic forms of the martial arts.

Today, a newer type of more practical "streetfighting" school has emerged, teaching the use of available weapons (knives, bottles, clubs, etc.,) as well as kicks, blows and holds.

The original Oriental techniques have been found in many cases to be too slow, involved, and complex. Experienced instructors now admit this, and have gleaned the best from the eastern arts and combined them with the graceless but lethal tactics of the alley brawler.

If dealing out pain with your hands is part of what you want to know, try a good Hand-to-Hand course. Expect lots of bruises, sore muscles, and injured pride.

SURVIVAL SCHOOLS

Field Survival schools have been with us for many years now, evolving from desert and mountain courses of instruction originally the domain of the sportsman.

Interest in "sports" survival is still with us, but it has been overtaken by "survivalists" who want to learn fieldcraft as a possible means of cheating the grim reaper the day after the apocalypse.

True field survival schools are dedicated to the great outdoors, and are actually the epitome of Boy Scout practicality. Fire building, shelters, finding food and water, land navigation, first aid, identification of plants and animals, and even skills like swimming, climbing, and hiking make up the schedule of a good fieldcraft school.

These schools have traditionally existed in the far west or northern United States, in scenic and rugged country, and had mountaineering or desert treks as their mainstay.

Fieldcraft is the least spectacular of all types of mercenary training. It doesn't involve guns, ambushes, or booby traps, and for that reason, is the most neglected of the group.

It is perhaps the most necessary. Don't substitute knowledge at making 100-yard offhand pistol shots for how to locate and purify water to drink. You'll spend more time in this life drinking water than you will shooting at people.

Field survival is the foundation of your skills as a field soldier.

MERCENARY SCHOOLS

Paramilitary schools are the newest schools of the lot.

They generally teach military skills in patrolling, ambush,

weapons, and explosives. This is the sort of school to consider if field combat is your goal.

I should make it clear the paramilitary school concept is controversial, and you should choose the one you wish to attend with some care.

Any school is only as good as its staff and course of instruction. Specialized schools such as hand-to-hand or shooting schools have it a bit easier in that their subjects are more limited, and it is easier to judge their performance.

A paramilitary school may incorporate many of the aspects of all the schools we've discussed so far. This is a big job, and takes experience, equipment, and time.

When you consider a school of this type, ask a few key questions, such as:

1. ***How long is the course?***
(The longer the better, as a rule of thumb.)
2. ***What are the qualifications and experience of the instructors?***
(You don't need to recognize their names, but they should be well based in the skill of their field.)
3. ***How long has the school been in business?***
(Start-up and failure rate of this type of school is high.)
4. ***What are the living conditions?***
(A field course should take place in the field, not out of a classroom or commuting from a motel to the field.)
5. ***How much practical work is done by the students?***
(The more hands-on time you get with weapons, patrolling time, ambush hours, the better.)

Many paramilitary schools have racial or political aspects, explaining some of the controversy surrounding their existence. Check this out with the school you choose.

The value of a field combat school of the paramilitary type is obvious. It gives people who have never had military training the opportunity to learn Ranger-type tactics, in addition to making training possible for those who may have been turned down by the military for health or physical reasons.

Independent schools can vary from military doctrine and teach subjects the military (for political reasons) cannot offer.

You don't want to sign on for a course at a paramilitary

school somewhere only to discover you've paid good money for some ex-military jock to DI you around with rock-laden pack runs or pushups. This sort of harassment is common in the military, but has no place in a private school.

If you're new at the Merc business, and have no idea of where to begin, the best place to start is at a security or combat-type school. You'll find people like yourself there, and some of them will have experience.

The training you'll get at a school will be the beginning of your career. Recruiters watch schools for good students.

And when that recruiter calls you, keep in mind what we've been talking about here.

I'd like to see you live to spend it.

INTERNATIONAL TRAVEL

When you're on your way to complete that contract, international travel becomes part of your job, and traveling involves dealing with different customs, people, laws and situations.

The jobs are far and wide, from the Mideast to Central America, and the quicker you can order ***chi*** (tea; served hot and sweet) with an arabic flourish, or ***cafe con leche*** (coffee with milk) like a spanish native, the easier it will be for you.

Home grown country boys might make it in a foreign country in the U.S. Marines where there are wiser people looking after them, but not wandering alone on the streets of some foreign city. Alone, you have to know what you're doing, or you will wind up broke, lost or worse.

Lets begin at the beginning—getting your passport, arranging for visas, selecting your transportation and buying tickets.

PASSPORT

Passports are necessary for international travel. They are a means of identifying you, and keeping tabs on where you've been and when. It is your license to travel, and it ***can*** be canceled or revoked if you sufficiently irritate the authorities.

To get a passport, go to the ***main*** post office in your city and ask for a passport application. It's a simple form, requiring only proof of your birth, current address, the fee, and a full face photograph that will be sealed into your passport.

It can take awhile (four to six weeks) for the government to issue your passport, so don't take on any urgent committments and expect to leave on your own scheduling.

LANGUAGE

While you're waiting for your passport to be mailed to you, there are a few things you can do to occupy your time.

Go to your favorite bookstore or library and find out all the information you can about the country where you're going. You want to study maps, history, statistics, photos, the lot. This is just basic intelligence, my friend. The more you know the better off you will be.

A basic study of the language is going to be essential. The people there are ***not*** going to be able to understand you when you get off the plane, and ability to say a few phrases is better than nothing. Reading train timetables or trying to arrange a hotel reservation when the language barrier is too high to see over, ceases to be amusing if you're late or have nowhere to sleep for the night.

Those foreigners are often so rude as to not speak anything but their own incomprehensible language, and you'll quickly discover, saying what you want them to understand slowly (or loudly) in English won't work.

If you're a fast learner, you'll enjoy catching on to another language. It'll teach you a lot about your own as you have to teach your assistant machine gunner the beauties of God's English and rack your brains trying to remember (in an emergency, of course) if "shoot them" or "shoot me" is "sa-we'do" or "Sa-we'da" in his own tongue.

The small pocket language guides sold in book shops are good to have. I take mine and tape the covers with green cloth tape to kill the reflection and to help protect the book itself in the field.

A word of caution; ***don't*** ever pretend you are fluent in a language in order to get a job or simply to impress someone. You will only end up looking like a fool, and fools are not paid to accomplish missions.

In America, we are sadly short on second languages. Unless you live in California, Texas or Florida, you are not often exposed to Spanish (New York Spanish ghettos excluded for clarity) on a daily basis. Since south of the border is going to be hot for a very long time, Spanish is an excellent choice if you are looking for a second language to study.

Learning a language is one thing in a classroom. It is something else if you are living with the people and dealing daily with their language. You will learn faster if you're undergoing the total "immersion" method.

VISA

A vital piece of information you'll need is if the subject country requires you to have a ***visa*** to enter and leave its borders.

Most countries require permission (as a visa) to get in and the really entertaining ones require permission to get out. Not to mention cloud-brained foreign clerks who smilingly take your passport (swearing to have it back for you with the visa in seven days) and it vanishes until next spring. Let your identification be lost like this (common problem) and you'll be sweating bullets every time you have to go through a roadblock or papers check.

Some countries issue their visa as a rubber stamp whacked without question directly into your passport at the airport or border station. Others make you mail your passport to their nearest representative embassy or consulate, where they carefully consider if they want you inside their country or not, making a decision based on your religion, military and police records, and so on. If they accept you, they stamp your passport with their visa, and specify the terms of the issue of the visa. This is usually the limits of the length of your stay, and any prohibitions on employment.

A select few tourist-oriented countries close to the U.S. require ***no*** passport from a U.S. citizen as long as you can show some proof of U.S. citizenship.

If you don't know if a visa is required, if it is issued through a consulate or embassy or at the point of entry, you'll have to write or call an embassy or consulate of the country you want to enter. Since all countries have embassies or consulate in Washington, D.C., simply get on the phone and ask long distance information in Washington for the phone number of the embassy you want, then call that embassy and ask. Most of them have information packets they will mail you outlining their requirements.

EMBASSY AND CONSULATE

While we're on the subject, what's the difference between an embassy and a consulate? An ***embassy*** is a full service concern. A ***consulate*** can provide most embassy functions (passport/visa service) but not all, because it is smaller, usually only a single office.

When affairs become strained between countries and embassies are closed, consulates are often left open as a line of communication even if the consulate office must be hosted in a neutral embassy for the protection of the consulate employees.

TRANSPORTATION

Transportation between countries can take more forms than flying. You can go by ship, boat, car or train. What means of transportation you choose is determined by these conditions:

1. What types of transportation serve the area you want.
2. The frequency and reliability of the transportation.
3. Cost of the transportation.

An important point is to make your transportation match your cover. You may want to fly into a neighboring country and then enter the target country by bus or boat, or you could see a reason to travel across an entire country by rail rather than spend your budget on airfare.

Remember public transportation abroad is often ***different*** from the United States. In Europe, the trains go everywhere, on time, night and day. In Africa, busses are a prime means of overland transport unless you're using river boats. The same is true in underdeveloped sections of the Mideast or Asia.

Don't take chances when you travel. Make ***sure*** the bus or cattle truck is going to where you want to go. It isn't funny to discover hundreds of miles too late, you're heading to some exotic spot not listed in your tour guide.

Ask questions and double check everything. Better to appear the fool by asking for the right train at the station than ***be*** the fool and have to jump off just before crossing a border you hadn't intended to encounter.

W.H.O. CARD

Besides your passport, you are going to need a shot record card. This is officially known as the World Health Organization (WHO) card, and it contains the list of your innoculations.

Some nations are picky about who they allow in, especially if they've had a particular disease under control and don't want to reintroduce it to their citizens again.

To get your WHO card, go to the country health center in your area (the card is free—the shots are not) and tell them where you're planning on traveling. They will know what to give you and in what doses. (The most fun of getting this document is having to take all the shots they want you to have in order to qualify for wherever it is you're going—"Nurse, give this man two injections for the Black Plague, one for Typhoid, and all the penicillin tablets he can carry.")

Many large airports are equipped with a health office and can give innoculations on the spot to travelers. This is good if you arrive and are told you don't enter the country without an update on your plague shots. Sometimes this service is free, or available for a modest fee.

A plastic or leather passport case is nice, since it keeps your passport clean, and also holds WHO cards, tourist cards, etc.

TRAVEL AGENTS

Now what about the price of all this traveling?

It costs money, and usually lots of it. The money will generally be yours, so you want to travel in the most economical way.

Don't just go to the air terminal and buy a ticket at the desk for a long flight somewhere. Let a travel agent do it for you. They can find the lowest fares and will help you with visas and current travel information about your target country. The use of a travel agent will save you time, trouble and money.

It does not cost you to use the service of an agent.

The agent can book you on group flights, make hotel reservations for you along your way with hotels who give them special rates, even arrange for guides or genuine tourist tours if your cover or intelligence gathering operations demands it.

You just don't have to tell your travel agent the whole story. (Yes, I've just always had this longing to visit beautiful El Salvador . . .)

Travel agents can come up with some oddball means of travel as well as the conventional. It isn't hard for them to place you as a passenger on a freighter, for instance, if said freighter has extra staterooms and sometimes accepts passengers. Freighter travel is slow and scenic (every port of call down the South American coast, for instance) but it is much cheaper than jetsetting if you need to save money.

It is best to make all arrangements for tickets, and reservations, ***before*** your trip. International travel is enough of a circus without making it more so. The unexpected will pop up as sure as bad apples (canceled flights/coups/weather problems/etc.) and little things that go wrong can blossom into ***big*** troubles for your operation.

This chapter is not intended to be used as legal advice. It is information only. Go to your library and look up Title 18, Statute 956 and Title 22, Statute 441, for the legal specifics of the Neutrality laws.

THE NEUTRALITY ACT

Merc work has it drawbacks, not the least of being the possibility of loss of U.S. citizenship. The fine print in your passport only hints at the revocation of your citizenship, and you'd be wise to go ahead and ask for details if you think part of your job might be saluting another flag.

You know about your opponents in the field, those desperate types with revolution on their minds and Soviet weapons in their hands. Now be aware you have another, less visible set of potential adversaries—men in business suits with law degrees.

You'll discover you have not only your contracted enemy, but the entire Justice Department/U.S. Government against you as well. Uncle Sam is funny. He thinks he has a monopoly on armies (large or small) and takes a dim view of those who try to play in the big league.

The Neutrality Act is not a bluff. It is a very real law that carries prescribed punishment—meaning years in a Federal lockup—for a citizen of the United States who is involved in combat actions against a nation with which the United States is not at war.

It is a law that is directed against mercenaries and any others who conduct (or in some cases ***conspire*** to conduct) any private or independent actions involving sabotage, combat, etc.

At first glance, it would seem the Neutrality Act would put a stop on merc activities. If it were rigidly enforced, this might be true.

Now we come to the loophole.

The U.S. government seems to pick and choose who gets the heat for Neutrality Act violations, and they do so with some selection. The Act is usually enforced against individuals who are running counter to U.S. interests, such as planning to attack a friendly government, or the Act is used to nail those who the U.S. government wants to get on other charges but cannot.

The Neutrality Act as we know it was enacted in 1948, with later updates.

Being investigated for violations of the Act is one thing. The Department of Justice chooses to actually prosecute only a few of the cases investigated. If arrest and prosecution is ordered, conviction is still rare.

The Neutrality Act contains many sections and specifications, but it is not exactly a precise law, being left open to interpretation by defense and prosecution alike. Actions taken by the defendant during the planning, organizing and launching of an alleged military attempt against a foreign government can be viewed by the deciding jury in whatever way the most persuasive attorney presents them.

There are many cases where American citizens have traveled abroad to join the army of a foreign nation. Rhodesia and South Africa come instantly to mind. Those Americans were not arrested.

Why?

First, the individuals who did it were discreet. They didn't advertise their activities, and while some U.S. intelligence personnel might have been aware of their presence, if the citizens in question keep a low profile (staying out of the news and out of embarrassing trouble like being captured by the Other Side) there was no reason for the United States Government to go to the time, expense, and effort to investigate and prosecute them.

Foreign armies, as a rule, are no more fun than the good old U.S. Army that you hated so much. They don't pay any better, either. Joining a foreign army is only good for exposing yourself to a foreign culture, perhaps getting some combat experience, and having someone feed and clothe you while you find out what kind of opportunities are really out there.

The Neutrality Act is a handy law used by the U.S. to get at those people or organizations who are making waves.

How do organizations like "mercenary schools" keep from running afoul of the Neutrality Act? The Act does say it is a violation to train, plan or conspire, for a ***specific*** mission against a legitimate, recognized government.

Training in general is okay under the Act. If you have a class in the field undergoing routine training in general subjects, even though you call them mercenaries, Uncle Sam hardly bats an eye.

If you had the same people rehearsing a raid claiming to be performed against the capitol of country XYZ, you ***might*** have a violation.

If those people can be proved to be purchasing weapons and equipment for that mission, the case is stronger. If those people can all be caught trying to leave the country for their target, bingo, you have something hard and fast with which to arrest and prosecute.

The FBI usually likes to wait to make arrests when the mission is actually being launched. They have the whole package then.

OFFICIAL SYMPATHY

If you keep your mission and all its trappings invisible, you don't make or cause trouble for the United States, you break as few laws as possible, and you are taking on a job against an unofficial guerilla group, and not a recognized nation, Uncle Sam just might look the other way. Sam's already got all the hassle he has budget for, and he'll go after those he can get, not those who make it very, very hard to be caught.

You might even luck out and find yourself doing something for which there is U.S. government or public sympathy. Recall Bo Gritz and his Loatian prisoner rescue attempt that went bad? He was slapped on the wrist, when there was every bit of evidence necessary to prove violation of the Neutrality Act and take it to a conviction in court—*if* the jury would have gone by the book and not by their hearts.

THE GOVERNMENT VIEWPOINT

Those men and women who get their paycheck from Uncle Sam have agreed to abide by and uphold the laws of their country. They have to pursue you if you violate those laws, regardless of their personal feelings about the validity of those laws.

Those FBI agents downtown are being paid for the same thing you are—to do the job.

There are some things for you to consider. What if the U.S. government is already conducting a covert operation in the area you intend to penetrate or raid?

What if your target isn't what you think it is? It might look, for all the world, like one more drug operation, but there could be inside people there who report to a CIA case officer, helping to locate and identify larger targets.

The point is, you don't always know if your mission isn't running contrary to U.S. interests, and the more your op threatens a solid undercover action, the more heat you're likely to draw off Uncle Sam.

The government generally looks at you as a pain in the ass in varying degrees, a wild card that has the potential to hurt as much as help. If they see ***help,*** they might pay you a friendly visit and establish some cooperation. If they see hurt, they will run you into the ground.

MERCENARY RECRUITING

The recruiter is the person who does the hiring for an operation or job. To be a merc, you have to start somewhere, and all missions begin with being hired.

We want to make several points about recruiters, so you can tell the honest from the dishonest, the professionals from the jokers, and a genuine job from a daydream.

Let's start with the types of recruiters. There's not that many, so it'll be easy.

I. ***Official Governmental Recruiter.***
II. ***Official Corporate Recruiter.***
III. ***Private Legal Recruiter.***
IV. ***Private Illegal Recruiter.***
V. ***Rip-off "Recruiter".***

I. OFFICIAL GOVERNMENTAL RECRUITER

An official Government Recruiter selects and contacts personnel for government-sponsored operations. The Official Government Recruiter is usually after recently discharged or retired military people, but sometimes approaches active-duty personnel. This sort of Recruiter is often an agent for an intelligence agency, like the CIA, FBI, or U.S. Army Counter-intelligence.

The Official Government Recruiter is the one many other types of recruiters attempt to imitate, to fool or mislead you into believing they represent "the government".

Official Government Recruiters hire for genuine government backed operations, and have the pull to get active duty military personnel changed in rank or other status, so they may be released to assist on a mission.

Private contact by your own (Read C-I-A) or another government for covert action is a sticky, double-dealing process at best. The best thing is it happens so rarely we don't have to worry.

Yes, even with all of our top-notch agents and out-of-sight intelligence budgets, sometimes the big boys have to stoop to doling out a little money and staffing a special "operation" to get something done without ending any government pension possibilities or seniority records by means of an enemy bullet. They carefully (and in the best clandestine ways) arrange, hire, and kick off some absolutely crazy missions. The good points are they almost always pay, the bad points are if you get stuck with that type of work, they'll probably keep you with it until you do get killed, and then claim they never heard of you.

To even be considered for this type of employment you have to have a file left behind in a government drawer when you were with the Special Forces. SEALs, Rangers, etc. They even provide names to friendly governments and let them hire you. If having your name leaked to Israeli intelligence is your idea of fun, and you get a job proposal that looks like an old *Mission Impossible* script, be careful. You can get into trouble just asking questions.

The Official Government Recruiter does approach civilians who have a required skill or type of experience, even if the civilian has no prior military service.

A big catch is, the Official Government Recruiter may sometimes come to you under cover of ***any of the other types of recruiters,*** in order to maintain secrecy for the actual employer . . . Uncle Sam himself.

II. OFFICIAL CORPORATE RECRUITER

The Official Corporate Recruiter is looking for people to carry guns, do deeds dirty and otherwise, and provide security in the form of investigations or bodyguards for the corporation.

Close attention is applied to international legal matters, and little lawbreaking is done. Corporations have armies of lawyers to bend the law or eclipse it when necessary, so permits and permission for the security personnel are usually there if sensitive items like firearms, espionage-type electronics, or fake documents are required.

Official Corporate Recruiters are often ex-Official Government Recruiters, and there is a certain amount of crossover in the people they headhunt.

Official Corporate Recruiters don't usually pass themselves off as anything else. Your paycheck will be coming from corporate headquarters like the rest of the employees, and you'll have the same benefits and perks as the office personnel.

Official Corporate Recruiters sometimes run ads to locate people for their security departments. Sometimes they don't, if they have a policy of keeping security to such a low profile as to be invisible.

Maybe you don't see the connection between "mercenary" and security people. It is hard to distinguish. They both accept money to provide protection, or to get a job done where force of arms is the decider. A pistol-carrying bodyguard is as much a merc as a rifle-carrying recon team member.

Being hired by an international firm for security work is very popular these days, because there's so much of it due to the razor-edge state of the world at large. Terrorists love to blow up banks, capture executives, ambush or attack corporate vehicles, and cause strikes or otherwise intimidate the local workers into not working, and therefore deny the capitalist warmongers of a few more dollars. All big international firms have security departments that often resemble small standing armies.

They pay well and demand good people for their money. Your letter of request for employment to Coca-Cola, Exxon, etc., should be professional and absolutely devoid of any James Bond-like leanings. They want well trained, tough, intelligent employees—and the key word here is ***employees.***

You'll be hired by the same apparatus that screens and hire secretaries, so keep a low profile while you're going through those first long steps to the final interview by that individual from security. After they hire you, you can freak and start wearing the shoulder holsters and learn to write your name with an Uzi on full-automatic.

III. THE *PRIVATE* LEGAL RECRUITER

The Private Legal Recruiter is someone who has the knowledge of who needs people for either business security, private security (bodyguards contracted to a single individual, for instance), or foreign government missions or jobs.

We designate the Private Legal Recruiter as ***legal*** because this recruiter does not seek personnel or hire for operations that are obviously illegal.

The Private Legal Recruiter gets paid for locating qualified people for qualified employers. This type of recruiter plays it straight with both sides, and because of reputation, becomes an excellent unofficial source of jobs for mercs and hired guns.

The Private ***Legal*** Recruiter has a direct counterpart you have to be aware of . . . and the difference is so slight, often you can be fooled easily.

IV. THE PRIVATE ILLEGAL RECRUITER

The Private Illegal Recruiter's description almost matches that of the Legal Recruiter, but with a single difference.

The Illegal Recruiter will take the task of providing people for missions or jobs that are flatly illegal. What I mean by illegal is this: ***any operation that breaks national or international law as a basic premise of its organi-***

zation, or needs to break such laws in order to be carried out.

Enlistment into a private strike force or security team is perhaps the most well-known but least active type of mercenary work there is.

These tend to be flatly illegal and hair-raising at their worst and comical blunders at best. Sometimes it's a combination. Private armies get started by someone who either has the money (and the nerve to impose his or her will by force or arms), or thinks they will get the money (if the coup/attack/threat succeeds).

Private armies or security forces are often not mercs at all—they are mob hit-men, engaged in gang warfare and nothing more. A word of friendly advice, don't take any contracts for "mercenary" work on United States soil. Here it's not covert operations, it's plain crime, and if you're tempted by a couple of thousand to go somewhere, and plant a bomb to a guy's car, you need your priorities examined.

Taking arms as a U.S. citizen against a country with which the United States is not at war is a violation of the U.S. Neutrality Act, and is illegal.

Hijacking an aircraft for the purpose of conducting a mission or for escaping from a mission is a violation of International Law. This includes ***military aircraft*** of all nations, making a scheme to hijack a Soviet HIND-D out of Nicaragua is an illegal act even in the eyes of the United States.

Murder or kidnapping of citizens anywhere in the world is illegal.

Crossing an international border without permission of the country involved is illegal.

Smuggling is illegal.

Etcetera, etcetera.

You might have raised your eyebrows after reading this, thinking this covers the majority of merc work. Maybe it does. We are not giving advice on morality here, just information on recruiters.

The Private Illegal Recruiter is more covert than the Private Legal sort, and may present a job to you without elaboration of the actual illegalities, assuming you don't or won't know the difference.

Whatever the failings of the Private Illegal Recruiter, they don't raise the ire of mercs like the lowest of the clan of recruiters, the ***Rip-off Recruiter.***

V. THE RIP-OFF RECRUITER

The Rip-off Recruiter comes in two forms. The intentional, devious, criminal recruiter who takes money from you for work and does not deliver, and the just plain dumb recruiter who books you into incredibly screwed-up operations where your best chances consist of getting killed or captured before you get so far along as to discover you won't be paid anyway.

The devious sort is nothing more than a confidence crook, who creates ficticious work, and then collects from the hopefuls who would go, telling them they have to pay a fee for their job, or some such other tale.

What makes it hard to detect this sort of con game is many very real jobs require you to get yourself there first, before you can be hired.

The con game recruiter preys on this reality and twists it to personal advantage.

If anyone attempts to pass themselves off as an Official Government Recruiter, it is the Rip-off Recruiter. Presenting themselves as an official ***anything*** is a way they have of strengthening their position with you.

Some Rip-off Recruiters have actually taken money from rebel groups, foreign governments, or private employers, then in turn taken money from the merc who were to be hired, and skipped out on the entire affair.

A common Rip-off Recruiter trick is to run ads for mercenary, military, or foreign employment, possibly even fronting it with an office, and stretch out the charade of "hiring" for as long as it will fly.

The only reason the Rip-off Recruiter can do as well as he usually does is because of the naturally secretive, covert nature of merc hiring. Real recruiters often can't reveal to you who the employer is, where the job is, or even what the job is, until much later in your relationship with them.

The Rip-off Recruiter has a con game tailor made, where half-truths and hinted-at connections are the rule and not the exception.

The other sort of Rip-off Recruiter, the dumb one, differs in that he is not intentionally trying to foul you up. Ego is the prime mover here, masked by patriotism, politics, or other hatreds.

Recruiting for mercenary work is a business that takes great depth of experience. The recruiter has to have the years and operations under his belt to know who and what is needed to get a job done, so the operation produces wealthier mercs, not broke survivors.

GETTING HIRED

A legitimate sequence of events for a merc job follows a pattern. It goes like so:

1. You are contacted by a friend of a friend, or you apply for an operation you know or suspect is going on.
2. The recruiter meets you, having first been given a fairly thorough run down on your background by someone close to him who he trusts, possibly a friend of yours who is already on the job.
3. You are not at first given working details and specifics of the job. It is even possible you may be given a cover story to conceal the actual mission. This is normal security.
4. Pay, traveling arrangements, expenses, and other points are agreed upon early. This is a business arrangement. Ask business questions.
5. Upon acceptance, you will be given deeper information about your part of the job, probably bit by bit as you begin to train or actually go to work.
6. A training session for any mission is a must. If you're told all of you are so experienced and so good you don't need to do it, trouble has reared its ugly head.
7. Some confusion and delay is normal, even on the best operations, so be tolerant when phone calls, expense money, or schedules get fouled up . . . but limit tolerance to good sense. Continual foul-ups are the kiss of death, and mean the employer, the plan, or the personnel are not right.
8. The professionalism and ability of your leadership people, and the experience and ability of the rank and file of the mission will give you a fast appraisal of the reality and chance for success of the operation. Each person's job has to match their level of competence. Specialities to meet the needs of the mission have to be part of the skills of the personnel. If the use of serious high explosives are planned

do you have an actual explosives expert, and not just somebody who's "done a little work" with explosives before? Keep your eyes and your mind open.

HIGH RISK

Jobs that have extremely stretched chances for success, like those *Mission Impossible* scripts, are either crazy or false. That timed-to-the-minute action only works on the silver screen. Don't bet your life on it.

On this count, you have to be honest with yourself. Since when has your luck been so good you can actually believe that with little training and almost total dependence on circumstances, you and a few other swash-bucklers can jaunt into Central Buga Buga and break someone out of a maximum security prison?

Rip-off recruiters hire for missions like this, playing on your desire for adventure as the lure. There really ***are*** missions of this sort, and they can sometimes be pulled off, but understand, they take rich resources and excellent people, not a rag-tag band of dupes that have paid the recruiter in advance for tickets and hotel accommodations to Buga Buga.

A real mission has at least a reasonable chance for success. Smart mercs don't take any other kind.

There is no established scale to weigh pay vs. danger, but a rule of thumb is the more hazardous a job the higher the pay will (or should) be. "Short term" work almost automatically means dangerous work.

Actual rates of pay are not as high as most people believe, because in the movies, mercs are always getting incredible sums to do incredible deeds.

A hot weekend raid—which might involve shooting—could pay no more than $5,000. ***Extreme*** risk (kidnap or assassination of a head of state, for instance, might pay two or three hundred thousand, not millions).

Use your judgement on pay scales. If extremely high rates are being quoted, be cautious. Most governments, corporations, and individual employers don't want to pay any more than they have to pay to get a job done. A lowball figure is more apt to be real than a six-digit number that promises to put you into Beverly Hills on completion of the contract.

Likewise, promises of sophisticated and expensive support (helicopters, boats, etc.,) for a mission that does not rate the expenditure is suspect. Rip-off recruiters often make James Bond-like scenarios as part of the sell to you.

There are a few requirements I like to have for a prospective employer or mission. I'll list them here, but don't think they necessarily appear in my order of preference.

1. Who is it for?
2. How dangerous is it?
3. How much does it pay? How do I collect?
4. What equipment and weapons are available?
5. What kind of support (moral/legal/artillery) will we get?

Number One is the biggie. there are some people I will not work for. This is a complicated decision, because there are times the "bad guys" can hire you for a good reason.

I wouldn't take a job paid for by Saudi Arabia against Israel, but I would take Saudi royals for a sortie against the Iranians . . . get the drift?

Number Two is really a loaded question. I suppose it gets down to how much you think your one and only life is really worth.

Number Three would seem to be the heart of the matter, but here again I'm flexible. If the right type of mission comes along, ***I'll*** pay to go on it. Pay is relative. $5,000 for a weekend on a very dangerous operation could be equal to the same amount spread out over five months in a soft billet somewhere watching the grass grow. Collecting is something else again. Does the employer actually have the money? What currency is it in? I don't want to get into any situation where I'm being paid in the 20th century equivalent of confederate money.

Number Four is practicality itself. Equipment and weapons get the job done. You can almost predict on some ops if you'll have a chance or not, based on what you have to do it with.

There's no percentage in fighting people who have automatic rifles if you're restricted to sharp sticks and good intentions.

Number Five is the trick question. Can the employer bail you out of jail or send in a platoon of commando types if your ass is to the wall? Find out the other assets of your boss. It's sad to say, but as a rule, if he had other assets, he wouldn't be hiring you. Something to think about.

SECURITY

DANGER SIGNALS

There are some signs you may have been compromised, or the operation is in trouble. Here they are:

1. Repeated requests for meetings by personnel involved in the planning or weapons/equipment procurement ***at their locations and their chosen times.***

This usually means you have come under surveillance. You might be walking into cameras and microphones.

Solution: You call the meetings in loud public places, keep your personnel guessing where they might see you next. Don't let them know until the last minute exactly where to go.

2. Pointed questions by planners or suppliers who seem to want ***you to commit yourself to an obviously illegal*** bit of merchandise they have or can get.

Solution: Ask suppliers what they can provide, what the prices and delivery details are for each item, then make your committment at the last minute. This prevents informer or surveillance personnel from being able to pinpoint your needs, or to have you on tape verbally agreeing to an illegal act.

3. Obviously loose security, becaue of the type of organization or individual who's requesting or funding the operation. Some political or religious groups freely gossip about activities to other "true believers" which is why most efforts by fringe groups fail. ***Remember. Most compromises are from the top down in an organization, not from the bottom up.***

ILLEGAL ADVICE

Pro mercs know how to operate around the Neutrality Act. Here are some of the rules they follow:

1. Don't plan anything inside the United States.
2. Screen and choose your friends and associates carefully to prevent infiltration by agents or informers.
3. Don't cluster your personnel together anywhere, or move them as a group. Filter them out of your staging country to the target by ones and twos.
4. Keep the personnel separate from the weapons and equipment. Don't let them know where it is until the time they are issued it for use on the operation.
5. Never disclose the operational target, plan, names of other personnel, or sources of supply to anyone in the group that does not have the need to know. This is basic security and keeps trouble down overall.
6. Keep vital information about the actual plan of operation secret even from your employer. Try not to disclose dates and specific movements. Employers tend to brag at cocktail parties or to casually tell wives or lovers ***your*** business.

THE INTERNATIONAL ARMS MARKET

(How to find, buy and smuggle weapons)

Buying a rifle or pistol for yourself from a friend or a local gunshop is one thing. Trying to buy 100 rifles and pistols, or 1,000 or 10,000, then moving them to your area of operations, is something else indeed.

This chapter is about how to purchase weapons by the thousands, and how to move them secretly across international borders.

THE BUYERS

That's you, or your employer, who has given you the

responsibility of locating enough weapons for a small army. Buying weapons takes three basic qualifications:

1. Money
2. Documentation
3. Means of Delivery

Of these, money is the most important. Items two and three can usually be arranged if the cash is there.

Documentation under legal circumstances means official authorization from the government of the country where the weapons are being delivered. This is what is called an "End User's" certificate.

Without an End User's certificate, no legal seller will sell or ship to you, because they know their firearms will be confiscated at the port of entry into your receiving country by the authorities.

To obtain a genuine End User's certificate for you to produce for the seller, you must first be legally licensed by the receiving country as an importer of firearms, and you must have the type of license necessary for the type of firearm or weapon you want to import. It usually takes a different class of license to buy a shotgun or pistol than it does for a machine gun or rocket launcher.

The U.S. State Department has a branch that handles requests to import, issues the End User's certificate and on the certificate it states what may be imported for that order. Each separate certificate covers only one specific order.

With variations, this is the practice in almost all free world countries. Having covered what is normal and legal, let us proceed to what isn't.

THE SELLERS

These are the people with the weapons. They come classified like so;

1. ***Weapons Manufacturers*** who are licensed and authorized by the government of their country.

2. ***Government Stocks*** of weapons that may be offered for sale to other governments or qualified (END USER) buyers.

3. ***Commercially Owned Stocks*** of military weapons (usually obsolete or surplus) that are offered to legally qualified buyers.

4. ***Illegal Stocks*** of weapons stored in areas controlled by private security or guerilla forces. An example of this would be a large cache of weapons owned by the guerillas in Northern Columbia, near their drug refining and processing stations which are so well protected even the Columbian Army can't hit them.

5. ***Semi-Official Armies*** who own stocks of weapons. These are normally rebels or forces of National Liberation, like the Front For A Democratic Nicaragua based in Honduras, or the Christian Militia in Lebanon.

6. ***Gunrunners*** who sell stolen or otherwise illegal stocks of weapons hidden away from both government and private military operations.

To buy from the legal sources, if you are not licensed requires some convolutions. Bribes don't work well here, because the governments are too big, and there are too many crosschecks . . . but there are ways of getting the job done.

Legal sources sell new weapons, first-line modern combat equipment, at the highest prices. Good examples are Colt Firearms, HK of West Germany, or FN of Belgium.

In order to get these types of weapons, you can use a second country friendly to your cause to act as a front for the receiving of the arms shipment.

Big arms deals involve big friends (and big enemies) so having the cooperation of a second country isn't unusual. As an example, let's say the country not only has political sympathy with your cause, but has vested interests (citizens, factories, money) in the probable combat zone.

This is true now of Panama, who has used their status to obtain arms for factions friendly to them in Cental America.

This second-country involvement is done often, the arms makers know it, but don't care as long as they have the money in hand and the paperwork on file.

Commercially-owned stocks of weapons are sold by arms "brokers" who have the right license to buy up government lots of obsolete weapons. Brokers can be dealt with differently from governments themselves.

Brokers make their living by ***selling*** weapons, meaning they have more than a casual interest in selling to you and they can sometimes find a way to do it that bends the rules.

A common method for them is to sell to you via second or third parties (that old trick again) which they have established long ago to handle "delicate" transactions.

For small operations, where a few hundred weapons are needed, as opposed to a country or political group that needs thousands, a broker has the authority to write off enough old Sten guns or Mauser rifles as inoperable junk and cover their trail with paperwork to prove all the weapons went to the furnace or crusher.

You pay a price for this service, since the broker is taking a chance. You'll probably get the weapons with the serial numbers removed.

Sometimes the ***semiofficial armies*** we mentioned earlier can be a source of weapons for you, as they sometimes sell off part of their charity-provided firearms to obtain money for other reasons.

This is common, and requires you to have a connection with them on the command level.

They usually have to sell off their weaponry secretly, to prevent offending the host supplier and perhaps getting cut off from more aid, so you don't have to worry too much about a leak from their end.

Gunrunners who have stocks of smuggled, stolen, or illegally purchased weapons are always in the woodwork, but they are the least secure to work with, as some of them are known for taking payment and not delivering.

They don't usually have large stocks of weapons, but they sometimes do come up with some interesting merchandise, being receivers of oddlot stolen military goods like night vision equipment, gas masks, crates of spare parts, and so on.

It would seem that mercenary operations would rely more on the gunrunner than any of the other sources available, and many merc operations do, but larger-scale ***successful*** operations can't, as the gunrunners are not reliable and they don't usually have a resupply capability to offer.

Gunrunners are frequently associated with any number of other illegal affairs, like drugs or the fencing of hot boats or aircraft. You'll find gunrunners in upscale social circles, lodged in expensive hotels rather than in dives, with legal investigators never far from them.

CHOOSE YOUR WEAPON

There is a trend among third-world nations to want current, modern weapons so their armies will look like the armies of the major powers.

Enough rifles and machine guns that small armies won't own out of vanity are rusting away in warehouses somewhere right now to make the price of Enfields or SKS's awfully cheap.

Don't ignore WWII rifles or second-generation submachine guns when you're shopping. A new weapons system may make an old one obsolete, but an obsolete .30-'06 rifle cartridge will kill you as dead as a current 5.56mm NATO.

AMMO, SPARE PARTS, MAINTENANCE

Old weapons or new, you've got to have parts and ammunition for them. If your armory is aged, the costs of large lots of ammunition will be cheap (a plus if the extra money can go into your pocket). In 1983, Lebanon made a deal with Guatemala to trade most of their WWII arms and ammo for staple food items—beans, rice and coffee.

The Lebs had tons of old Nazi, U.S., British and French small arms ammunition, plus both workable and junked weapons for most of it.

One boatload of the Lebanese tradegoods made the error of docking at a U.S. port on its way to Guatemala, and the BATF confiscated the entire cargo.

Those unopened cans of 40-year old ammo were still effective, with the U.S. and German ammo seeming to be more stable than the British, which used a different type of propellant.

Age causes a change in the burning of small arms "powders", often making an explosion out of what should be a controlled expansion rate, damaging weapons and possibly injuring the users.

Primers die of natural causes, with a greater number of hangfires and misfires in unpackaged ammo that has been exposed to weather and handling.

Civilian ammunition isn't as well waterproofed as military spec stuff, nor is it packaged right for long term storage, ***so avoid old lots of commercial ammunition.***

Armorers need tools, a place to work, manuals and spare parts. A percentage of the rifles or other weapons in your unit will be needing repair all the time because of wear, combat damage and general abuse.

There are still spare parts available for old weapons, and parts for new ones are rolling off the assembly lines everyday, but to be hoenst, your best bet will be to cannibalize since even well organized armies who have computers and unobstructed lines of supply can't seem to keep their spare stocks updated.

This means you'll need extra weapons for parts, so buy junker spares that don't shoot, or deliberately pad your weapon's needs so you'll have spares.

ANTI-RIP-OFF TECHNIQUES

Serious problems in merc arms deals can arise, so you should know what the dangers are.

1. No merchandise delivered to you after you make payment.
2. Old merchandise delivered in place of new.
3. Incomplete shipments.
4. Sabotaged or booby trapped merchandise (enemy action).
5. Armed ambush at transfer site (enemy action).
6. Arrest attempt by authorities at transfer site.

There has to be some trust in any deal, but when there is a minimum of that precious commodity, you can always arrange a simple payment-on-delivery agreement, where you hand over the money when they produce the arms.

Inspection of the weapons is vital to see if you're really getting what you're paying for, and don't just handle one sample out of a preopened crate, take the time and trouble to do it right.

If you're being rushed or otherwise denied the opportunity to check your purchase, something is probably wrong.

Armed friends nearby are handy, since the point-of-transfer is a very vulnerable place for you and ***the*** seller if trouble is afoot.

Armed ambush from rivals, or arrest attempts by police agencies usually comes when you, the money, and the weapons are in one place. The danger exists until the weapons are relocated to a safe area.

A basic security plan for any point-of-transfer is to watch it closely until you're reasonably sure there's no waiting intervention, then move in.

If the sellers waiting for you aren't the people you expect, if there is no one there at all, or if the place just ***smells*** wrong, back out fast.

Speed, organization, and firepower break ambushes and raids, but that's the subject of another chapter.

To have any influence at all over the seller, you must be able to reach him ***after*** the sale and he has to ***know*** this.

Honest sellers usually make after-sale contact with themselves easy. A proposal made on a weapons deal between some independents who had large quantities of Red Chinese munitions and arms and a southern African rebel faction was to have the seller himself stay with the rebels as a voluntary hostage until the shipment was made. That's honesty.

CONTROLLED ITEMS AND SMUGGLING

Weapons are not the only restricted or licensed items mercenary groups need with which to fight their wars. The list below includes some, but not all, sensitive licensed or otherwise controlled items that can't just whiz through customs in most countries.

1. Radio transmitters
2. Night vision equipment
3. Ballistic (bulletproof) vests
4. Weapons parts
5. Ammunition components

The exceptions to "obvious" military items are aircraft or ships that have a dual business or military application. Actual armored vehicles, such as armored cars, are classified as if they were tanks by most nations, with some exception in firefighting or rescue configurations.

"Smuggling" can take the form of relabeling items to list them on customs declaration forms as something else—such as bulletproof vests becoming "protective clothing," or radio transmitters being radio ***transducers.***

Both are changes in nomenclature that satisfy customs agents by giving an accepted alternate, but technically accurate description of the shipped item. You have to be sure of the validity of your new description before using it. The best places to find out what customs misnomers are accepted by customs are the makers of the item themselves, who often use the same trick.

If your smuggling is of the nature that requires outright lying, like labeling crates of rifles as machine parts, you have two possibilities.

The first is to find a secure means of slipping your weapons or controlled items through customs by concealing them inside false bottom crates or inside other merchandise.

A large number of pistols being smuggled to Nigeria were welded inside water tanks, and flat packed inside spaced-bottom crates. It worked, even with the authorities looking for suspicious marks or evidence on that shipment.

The second method is to bribe shipping and customs officials. The shippers can include your load with a larger cargo of their own that will pass customs uninspected.

Customs agents are government employees and risky to deal with, so good advance investigation of the chosen agent(s) is in order.

Crooked customs agents routinely share such graft with their superiors, which means other people have to know about your shipment, leading to potential trouble.

A severe tactic to insure secrecy is to detail a hit team to keep the customs agent "bought" with clear and specific action ready for him that he ***is*** aware of, in the event of any actions against you.

Classical smuggling doesn't go through customs, but involves the delivery of contraband on deserted beaches or roads, usually under tense circumstances.

If you opt for a midnight boatload delivery, it is still dangerous but far more simple than bribing officials.

For the best security, don't hire local talent to do your transporting. Use your own boat or vehicles, and your own people. A guide familiar with your target area could be necessary, but it should be someone over which you have control. Assign him an escort who carries a silenced pistol.

Use recon personnel who have signal or radio commo with you to keep watch over all approaches, so your offloading and pickup procedures won't be surprised.

Adequate planning for interception while you're delivering requires imagination. Water delivery has the advantage of your being able to sink the weapons and ammunition in sealed cases just offshore for later recovery.

Offloading from an aircraft or truck has less options, and ditching the load and running may be out of the question. Armed force is the first means of defense you have if caught in an open, exposed situation.

In general, if you are using vehicles for final delivery to the storage site, plan out a route that avoids stops, delays and ambush sites, with no transfers or changeovers. Have your security teams in place before you arrive, with an established operating plan, fields of fire and communications.

CACHE, CAMOUFLAGE AND STORAGE OF WEAPONS

Weapons have to be hidden or stored until they are needed. Common storage in a secured area is uncomplicated. The weapons get lubricated, locked up, guarded and frequently inspected. The trouble comes when you have to do this in an ***unsecure*** area.

How you hide your shipment depends on the situation, but there are some basic considerations:

1. Prestorage preparation (maintenance/lube/packaging).
2. Decision to concentrate or disperse caches.
3. Guards to watch secret caches.
4. Action to take if caches are discovered.

Weapons can be left in warehouses disguised as other merchandise, and hidden inside huts or shacks, with your guards living around them. Caves, basements, culverts, even covered ditches are useful, anything that will keep the weapons out of the weather.

Well packaged weapons won't suffer even for long periods of storage, meaning a year or more.

You might decide to split up your weapons, ammunition, and accessory equipment into different caches so discovery of one won't expose to enemy intelligence groups all the types of arms, equipment and ammunition you have.

Think for a moment . . . the loss of a cache of your ammunition that contained all the types of ammo for all the types of weapons you have reveals a comprehensive picture of your operation.

It's more difficult watching many small caches than one or two large ones, but you knew the job was difficult when you took it.

BUILDING YOUR OWN

Actually manufacturing weapons isn't as wild an idea as it sounds.

A good 9mm submachine gun can be made using commercial steel tubing, steel bar stock, and the employ of a professional gunsmith.

One operation in Africa required the capability of portable machine shops on heavy trucks, a local team of experienced workers, two American technicians and their merc bodyguards.

The trucks could stay in one area until it was too hot (payoffs to village chiefs are only reliable for so long).

If your operation is low-level enough to employ zip-guns or other improvised weaponry, you can establish covert shops to make simple single-shot pistols and shotguns. All you need is experience, hand tools, and raw materials like pipe, wood, springs and fasteners.

LINES OF SUPPLY

Obtaining weapons is the first part of the chore. Keeping them fed is the rest. Ammunition for resupply has to be purchased, moved, stored and distributed. That takes time, and it's too late in the heat of an operation to discover you need more ammo and you don't have it.

Resupply, steady streams of ammunition, has to be worked into the plans from the start.

A single source might dry up. Arrange for two or three suppliers, and cache in strategic lots of ammunition ***and replacement weapons*** so you'll have a backup capability.

If it is at all possible, try to have your weapons shoot the same ammo as your enemy, so resupply can come from captured stocks. It cuts down on logistics, and there's something satisfying about throwing a little of your competitor's lead back at him.

MERCENARY LOGISTICS OFFICER

This chapter hasn't been about combat. It's been about ***logistics,*** which is the management of material.

There's more to war than gunfire. The successful merc, working his (or her) way to the top does it by assuming responsibility in planning, finances, command and personnel.

The merc on the line has to get his weapon and ammo somehow, and employment for the paperpushing end of merc work can pay better than the combat type who runs through the jungle night and day.

FALSE DOCUMENTS, FORGERY AND SMUGGLING

Merc ops often require the use of documentation made to order for the occasion, or the shipment of restricted or prohibited items across international boundaries.

In plain language, this means false documents, forgery and smuggling.

We'll deal with false documents first. A ***false document*** is technically different from a forgery in that you can obtain a false document by applying through routine and legal channels for it but with ***false information.***

The document issued to you then will be quite proper as far as stamps, signature, etc., go . . . but very false as to accurate information about you.

FALSE PASSPORTS

Getting a false passport is easier than you might believe. Proof of identification is the key here. When you examine the U.S. passport application, you won't see it, but there is a provision for you to provide ID for yourself outside of the "required" birth certificate.

FIRST METHOD

This alternate method of ID for the applicant is necessary. In many cases there are no records of birth in both large cities and small towns, so the word of a relative to the postmaster, as a signed and notarized document, saying you are whomever you want to be is deemed sufficient by the government.

A good example of this technique is to go to a small town (where people know each other) and bribe someone in a long-standing family there to swear to the postmaster you are his distant cousin who lost both parents while still an infant—and was brought up by an illiterate aunt back in the hills.

You can change the story to suit your own needs, but you see the basics necessary.

With a known citizen backing you this way, all you have to do is submit your application and photos to whichever passport processing center is appropriate to your area. Your photos, of course, should suit the cover you want to use, with hair dyed, beard shaved or displayed, glasses on or off, etc.

The false name you choose for this passport will be signed to the back of the photos just as the passport application requires.

When the passport comes to you in the mail, you're someone else.

SECOND METHOD

This technique is easier than the first.

Find a drunk or other dead-end individual who is about your age, and make friends with them.

Go to your local county health office and obtain a blank Birth Certificate form which will tell you what information you'll need from this drunk. The basic information will be simple, like names of mother and father, place of birth, etc.

Take this completed form back to the health office, pay the fee (usually $5), and they'll give you a copy of the drunk's birth certificate.

With the birth certificate, you now begin to assume the identify of the drunk. Build a background identity first—pay rent, utilities and other bills in the drunk's name. Now you're armed with enough background to go and obtain a driver's license in that name.

Don't worry about the drunk possibly having a driver's license already. You're using a different address. There are many people with the same name.

With the birth certificate, bills and driver's license, get photos of yourself and go and get a passport.

FALSE DRIVER'S LICENSE

Most adults are expected to be renewing a license, not applying for one for the first time. When you go to get your license, odds are they'll ask you for a previous license or license number.

You can fake it, of course, and take the position you are indeed getting a license for the first time, and your driving test should reflect this lack of experience.

You will be asked for proof of identification for this license, and not just a birth certificate. In many places they will want to know your resident address with some way of verifying it. This is usually a bill (power, water, phone) with your fake address on it, or official government letter to you at that address.

Yes, this can require an apartment you've set up strictly for a cover in order to use the mailing address. It may also be necessary to fill local residence requirements*.

** Some states want proof of how long you've lived in their state, asking for dated documents (rent, for instance), or a statement by your employer.*

If you have your new false passport, it can serve for the license personnel's requires for proof of ID, or you can actually use the same local citizen proof-of-identity vouch that got you the passport. This would work by doing them in the same town.

BACK-UP DOCUMENTS

Supporting identity documents that make up the common set of paperwork that carries your name should not be overlooked. With your passport and driver's license, you'll be able to get a library card, auto insurance card and club or membership cards. This gives depth to a false identity.

ALTERNATE METHODS

This won't work well inside the U.S., but you can put together an impressive display of documents that will look very good in Africa, Central America, or Southeast Asia.

The classified section of magazines of the paramilitary sort have ads by people who say they can provide blank marriage licenses, high school diplomas, driver's license and "ID" cards.

Of course, these mail order documents are weak, not having the capability of doing any more within the U.S. than fooling the careless, but they look ***great*** in West Africa.

If your document requirements can be filled with mail order papers, go ahead and establish a shallow identity with them. It is a cheap and simple way to fool someone who probably does not have much understanding of English, and does not know what the driver's license from your state looks like anyway.

FORGERY

Forgery is making or changing documents.

This is more work than applying for false documents and does not produce as solid a document. Forgery requires skill and imagination, and should not be tried by those who can't use cameras or don't have an artistic background.

Keep in mind a forgery will not stand up to a cross check if the official with your forgery goes to the trouble to call or cable. Customs and Immigration catches people who are trying to travel or pass borders with forged documents all the time.

The best documents to forge are those that are normally encased in plastic. This includes most U.S. driver's license, ID cards for the military or private companies, certain types of gate passes, and so on . . . actually, anything card or wallet sized.

By having the forgery sealed in plastic, the weight and feel of the paper or card stock is not apparent, so a different type of material may be used—such as a photo of a card rather than the card itself.

Polaroid-process cameras can make some beautiful copies of card-sized documents, but don't think you can simply buy a cheap Sureshot and go into the forgery business. It takes good cameras and lenses, proper setup, and slick art and lettering changes on the original to be copied for the finished photo to be believable. A professional photographer can do all this for you, and of course, you pay for it. Somewhat easier is having a print shop do your forgery for you.

If you need a letterhead (either a copy of a real letterhead or one you want made up) a print shop can do it for you. They simply photograph it and have it retouched by their staff artist/negative stripper, they print excellent copies.

This means you'll have to have a hundred or so letterheads run off in order to get the one or two you actually need, but war is hell . . . and often expensive.

The letterhead trick can get you any type of letterhead from corporate/legal stationery all the way to government agency or military unit letterheads.

A professional, well typed letter on the proper letterhead can get you many things both within the U.S. and outside. A caution though, if you're actually mailing these fakes, the receiver of your letter will reply to it according to the address on the letterhead, and that means it'll go to those people whose name and identity you've used. An official "government" agency letter to a foreign individual or company modified with the instructions to reply only to Smalltown, USA, Private Residence, will be met with much suspicion.

Some types of cards (black ink on white paper, for instance) can be very nicely forged.

A surefire trick is to have a photographer shoot the original to make an enlarged negative. A good retoucher can then take the negative and eliminate the original signatures, typed dates, and so on—and deliver back to the printer a negative ready to be reduced back to original size, transferred to a printing plate, and run off in whatever quantity you want. You then take the fake cards, fill them out with your information, and put them to use.

It's obvious the effect cameras are having on the forgery business. The counterfeiting of money, bonds, stocks, etc., is so easy now with advanced photographic techniques the only way the government or banks can protect such documents is to use special inks and papers the counterfeiters cannot get.

Matching papers is not as hard as it might seem if you have a project that requires duplication of an unusual paper stock. Here again the local printer is your key. Large printing houses have thousands of papers in all textures, colors and coatings.

Keep in mind a paper common in Europe or Africa might not be common in the U.S., but if you keep trying, you'll be able to get it if only by special order. Expect to pay well.

Magnetic or odd color inks are another security hurdle. Ink ***colors*** can be eventually matched if by no other way but mixing, but inks treated with computer sensitive metallic compounds are almost impossible to obtain. Better to try to fool the eye, not the machine.

Don't forget to add the ***authenticity of mistakes*** on your forged documents. Study the real thing and see how common strike overs, erasures, misspelled words, hand-inked corrections and smudges are. Don't overload a document with them, but use them as spice.

Perhaps the most simple method of forgery is to sit down with a typewriter and dash off a fake letter, adorn it with dates and signatures, and use it to impress border guards, inspectors and customs agents. You can do this on letterhead stationery or on plain bond paper. Simple letters like this can be giving you authorization to have or use cameras, recording devices, an automobile, whatever.

This is the most astute form of bluff, but bluff is your game. The SAS does say, ***Who Dares Wins.***

SMUGGLING

Books could be written on the fine art of smuggling, and we're not going to begin to be complete in a single article, so we'll confine this chapter to what you might be trying to take across the border with you.

The most smuggled items for border crosses are:

1. Drugs
2. Weapons and Ammunition
3. Money or Gold
4. Manuscripts, Maps, Papers, Books
5. Cameras, Recording Equipment, Film and Tapes

Let's take these in order.

DRUGS

If you're moving drugs, you're on your own.

WEAPONS AND AMMUNITION

If you're flying, and you don't expect your checked baggage to be searched too well, break the firearm(s) down and conceal them with other metallic objects, or hide the parts with unlikely things (.22 rounds inside a full jar of applesauce, for instance), weapons barrels can go inside sports equipment groups, receivers (surprisingly light after full disassembly) can be taped inside stereo players—use your imagination. Entire ready to shoot pistols can be packed inside hollowed books (an old but very usable trick), and ammo won't rattle if you tape it together machine gun belt fashion for rolling away and placing in a can or securing under suitcase liners.

Actually, don't go into a place with a firearm if you can help it, or carry a very small one ***well*** hidden, and use it to get a better weapon later.

MONEY OR GOLD

Currency is easily concealed, thank goodness. Try and get

your money into the largest demoninations of bills you can to make the job easier, then slit the seam of a plastic-backed notebook or writing board and stuff the bills under the plastic. A notebook (padded vinyl tape) can be in your hands without attracting undue suspicion. Money can be sewn into the lining of coats or jackets and never found on even strict, shakedown inspections.

If you're slitting seams or sewing linings, do it in a place not easily visible.

Gold is heavy and usually bulky if you have any quantity of it, so try to carry it in small pieces, spread out through your clothing and luggage.

I advise gold coins and proof-marked bars (1/2 ounce, 1/4 ounce, etc.) because the value of jewelry is not easy to determine if you're going to resell it to anyone but a jeweler.

Sew the gold pieces into your clothes in places the clothing would naturally hang. Don't put too much in any single garment because it'll get heavy fast.

Small pieces of gold are very easy to conceal in personal effects, especially in women's handbags, wallets, shoes and so on.

I keep recommending clothing for one reason; if your Escape and Evasion money or operations currency is vital, you'll have to have it close to hand. Smuggler's bags are subject to the same wild world of being lost, stolen, or delayed that affects the Idaho school teacher's luggage.

MANUSCRIPTS, ETC.

Manuscripts, Maps, Papers and Books can be hot items. These are generally intelligence items, and you're likely to find yourself trying to get them out, not in, to a foreign country.

You can have books recovered to appear to be a commercial best seller, by removing the orignal cover and having the cover from a similar-sized book glued on.

Maps should be folded and placed under false bottoms (the thin, shallow type, suitable only for hiding money or other paper items) in trunks, suitcases, or shipping boxes. A map page might be quickly placed inside a newspaper if you are in a hurry, or wrapped and taped to your back under your shirt.

Keep in mind, customs inspectors catch most smugglers because of the nervousness of the suspect, not because they have any idea there might be something illegal happening.

Written manuscripts can be labeled as something they are not, broken down and spread into envelopes as personal letters.

Manuscript pages on 35mm slides are easier to hide than the originals, so consider having a sensitive manuscript photographed by a good photographer using a tripod and copy stand.

CAMERAS, RECORDING EQUIPMENT, FILM AND TAPES

Since we're already on the subject of film, do get yourself a shielded bag for the transportation of film rolls. Airport "X" ray machines can and do ruin film, either exposed or unexposed.

If you can't get a shielded bag and you have to move the film, protect it with several layers of aluminum foil. You can carry the foil in with you in a "household" kit of picnic type accessories, so anyone asking about it can be shown plastic spoons and forks as well. Camouflage is very much the ally of the smuggler, so don't forget to make your cover as complete as you can.

Magnetic recording tapes are getting less bulky all the time. Carrying cassettes, if not too many of them, is no problem. They are small and light and conceal very well. If space is really a problem, don't forget you can break the plastic cassette apart, take out the tape, secure it with adhesive tape so it won't unroll, and make a far smaller package to drop inside the lid of a deodorant bottle, or whatever.

Cameras and tape recorders are in the age of true miniaturization now, and "spy" type examples of either sort are commercially available to anyone.

Those special cameras that use 8mm or 16mm film are to be avoided if at all possible because of the problem of finding film and processing for them. Like 9mm Parabellum ammunition, 35mm film is the international form, and you'll be pleased to discover very small, but very good, cameras exist now that use full-frame 35mm.

Small cameras don't excite soldiers and less well educated people like the big, expensive models. They take them for toys, reasoning no professional journalist or spy would carry or use something so small and unsophisticated as to be "obviously" ineffective.

Ha.

The super small tape recorders that use the tiny "micro" cassettes have the disadvantage of being restricted to that small tape and you'll play hell trying to find such a tape in Outer Bongo, but standard cassette tapes for tape players are found almost everywhere around this globe.

Carrying either small cameras or recorders is best done by doing it rather openly. Put them in your pocket if you think it safe enough, unless you go in knowing the ownership of such objects is illegal.

If you must hide them for cross border movement, the small size and light weight of these devices works to your advantage. Often, placing a small camera inside the inner pocket of a packed garment in your luggage is sufficient for a casual check, but restricted countries (Soviet Union, for instance) may be making very complete teardowns of everything you have.

This sort of empty-it-all-out on the table search makes the old smuggler's trick of specially constructed luggage or personal items a necessity.

Talc or lotion bottles (the plastic kind) can be made to house small recorders or cameras, with the seam made by cutting the bottle open concealed by the label. Again, don't make something too heavy. Weight is always a quick giveaway.

Any bottle used to hide an item should be still full of its original contents, so wrap the device watertight in a plastic bag, and refill the lotion or whatever—and rig it so nothing shifts or rattles inside.

Clever coverings for cameras or recorders can be used that are easy to remove. Knitted stocking covers that match a set of your ski hats or gloves will allow a tape recorder to travel in with the set.

Leather smoking accessories, boxed as a set, can have a camera inside the cigarette pouches. Use your imagination.

SMUGGLER'S NOTES:

The rules for not getting caught are these:

A. Don't carry anything you don't really have to carry. No sense getting compromised for an unnecessary chance.

B. Don't be nervous, talk too much, be hostile to the guards, or try to divert their attention from a part of your luggage.

C. Leave concealed items concealed until you need them. Don't go to the hotel and unpack it all.

D. Go out of the country the same way you came in with everything packed just right. If they have their eye on you, they might want to know what happened to the Ghetto-blaster stereo brought in for your third-removed cousin, or ask for an explanation of the fat panda dolls you're taking home to the kiddies. A clean-looking, well packed kit can subdue suspicion.

E. Everything must be accessible for inspection. Don't tape closed or lock shut something in hopes they won't make you unwrap it. They will.

If you try to smuggle automatic weapons somewhere, forge documents, or run an ID scam and get caught . . . remember, I never told you a thing.

FOOD

And food! Let's not forget food.

Are you the type who'd send a fine restaurant meal back because you found a rat turd on the veal? What in the world would you do if they furnished you with the whole rat?

McDonald's are few and far between in the Congo or lost in Saudi Arabia's terrible Rub Al Kahli desert. African and Central American cooking hygience is usually pretty casual even in the better spots.

American C-rations are not issue with squalid little armies. They tend to live off the land, or at best, out of cans printed with labels you can't read.

LET'S BE CAREFUL OUT THERE

On a merc operation, where you only have enough people or else to do the job, and you don't get paid if you fail, giving careful attention to the food you eat and the water you drink, can make the difference in a successful mission or a dud.

Plan for food and water with the same concern you do over your weapon and ammunition.

Rebel groups the world over are sustained not on packaged rations, but on vegetables from local farms and gardens, some livestock if they are lucky, or on the cheapest sacked food in the world, dried beans or rice.

The mercenary usually feeds himself off the local economy, taking fat times with the lean, and does not often have a mess officer or company commander to rely on—or blame—for the quality or quantity of the chow.

Of course, not all merc jobs happen in the jungle, or even in the field. Sometimes, they involve hotels and room service, or grubby roadside and street food vendors.

HOTEL FOOD

In the better quality hotels around the world, both the food and water is usually compatible with the tastes and digestions of the tourist trade. Many hotels actually pipe purified water to your room (a mirror sticker in the bathroom near the faucet will advertise the fact), and have clean, efficient kitchens as good, if not better, than hotels and restaurants in the States.

If your mission and budget allows for this class of hotel, by all means use them, and your concerns about possible stomach problems will be greatly reduced.

STOMACH TRAINING

A misconception about eating foreign food is that it is going to be dirty or contaminated and give a case of stomach and bowel attacks that will have you strapped to a toilet living off Kaopectate malts.

The truth is the cause of most traveling-related stomach disruptions is simply a change in bacteria. You are immune to the bacteria in your home area of the world. When you travel, you get doses of foreign bacteria in your food, your system can't cope with it, and you get sick.

You can immunize yourself to the bacteria in the locale where you'll be working by eating or drinking only small amounts of the food and water for a few days. Introduce your digestion to the bacteria in short doses, allowing your system to become familiar with the bacteria, and develop a tolerance.

During the meantime, only drink bottled water, and eat packaged foods. This technique usually takes just a few days, about the same period of time you would have gotten and stayed sick, and is far more pleasant than the alternative.

AQUA PURIFICADA

Very common in foreign countries, is the water section of the grocery. Bottled water of varying brands and types is sold under the classifications of purified or mineral waters.

The difference is purified water can be simply boiled or treated water, where mineral water should contain healthful minerals as well as being pure. Mineral water usually tastes better than purified water, but this is a matter of opinion.

If you anticipate travel where water is going to be a problem, even if you're not going to be in the field, go ahead and carry a means to purify or treat your own drinking water.

U.S. military issue iodine tablets are extremely powerful and will kill almost any kind of bacteria in water, while at the same time tinting the water iodine pink, giving it a distinctive iodine taste, and after a few liters of tablet-treated water has gone through your system, it even begins to purify the necessary bacteria that lives in your intestines, creating bowel trouble on its own.

If you can't get issue iodine tablets, you can use iodine crystals (available from pharmacies), or in an emergency, even tincture of iodine dripped right out of the bottle. Use about three drops per liter.

Halozone water purification tablets are not as powerful as iodine, but should do the job on hotel or restaurant water, without the taste problem. Sportsmen and campers in the U.S., prefer halozone to iodine for all but trying to purify water out of septic tanks.

Drinking local beer or soft drinks is a good but not a surefire way to avoid stomach problems, since bacteria is part of the making of beer, and it is certainly present in sweetened soft drinks. The bottles those drinks come in are not guaranteed to be sterilized before filling (or refilling) either.

COOKING YOUR OWN FOOD

It is not a bad idea to prepare your own food for extended hotel stays. Especially when the local delicacies threaten to cripple more of your team than enemy action.

With a small, safely used campstove in your hotel room, you can boil water, heat canned food, and generally cook almost anything for yourself.

FOREIGN LANGUAGE MENUS

Yes, some hotels and restaurants abroad have English menus, or menus in the local language plus English, but most don't. You ***must*** either learn enough of the local lingo to order food or you should carry a common tourist language guidebook that has a restaurant section in it.

All the old jokes you've seen about ordering bizarre local dishes or insulting waiters won't seem so funny after you've done it yourself, when all you wanted was something to eat without calling undue attention to yourself.

MEDICINES

Most cases of diarrhea are caused by ***E. coli*** bacteria. The drug doxycycline, part of the tetracycline family, may be used as a treatment or preventative.

Pepto-Bismol is great, as is Septra or Bactrim. Kaopectate works to limit and heal diarrhea, but ***after*** the fact. Travel with these medicines, don't wait for a member of your group to get sick and then go searching after dark for a late night pharmacy in East Beruit, downtown San Salvador, or wherever it is you take your "vacations."

By the way, the ***pharmacy,*** meaning drugstore, is common in many languages, usually altered only in spelling, such as ***pharmacie, farmacia,*** and so on.

FIELD FOOD

There are packaged rations sometimes issued to foreign armies, and sometimes even rebel or guerrilla groups. These rations are not what you might expect. Packaged rations are made to the taste of the region where they are expected to be distributed.

Dietary laws of some religious groups dictate what goes into "indiginious" (fancy word for local) rations, so expect Moslem rations to have no pork, or be cooked without the use of lard, or shortening derived from pork.

Jewish rations have to be kosher, Hindu rations won't contain beef. Knowing what is and is not tabu for human consumption in your area of operations can save you a lot of trouble. Chowing down on a can of pork and beans while on an Arab patrol rest break won't make you any friends.

U.S. military rations are sometimes part of guerrilla wars if the food has been captured or provided. Because U.S. rations contain a variety of foods either disapproved of, or downright disgusting to some foreign soldiers, be cautious as to what you offer to whom.

American WWII field rations were known as K-rations (names after Ansel Keyes, the nutritionist that developed them). The newer C (Combat) Ration was issued in Korea as stocks of K's gave out.

Both ***K*** and ***C*** rations were basically heavy canned foods with supplement packages of condiments (sugar, salt, etc.), packaged in cardboard boxes. Vietnam-era LRP (Long Range Patrol) rations are plastic-bagged, and are freezedried, and while extremely light, require lots of water (not always handy) for preparation.

The newest U.S. ration, the MRE (Meal, Ready to Eat) is packed in aluminum and plastic bags like the LRRP ration, but does not require water, making it only slightly heavier than an LRP ration.

NUTRITION

Long term field operations on nutritionally inadequate food ***will*** break down your body and your mind.

By allowing yourself to exist on a seriously deficient diet, you are almost guaranteeing the failure of your mission as surely as if you were working for the enemy.

The human body, reasonably fed, has a limited amount of time it can continue to function before running out of fat reserve, followed with health problems caused by lack of vitamins.

The fat reserves will be absorbed by the body when the stomach stops receiving food. Fat is energy. Loss of body fat reserves is demonstrated by a marked weight drop, and breakdown in strength.

Don't confuse a ***temporary*** drop in blood sugars and electrolytes which will cause a temporary loss of energy, with the ***permanent*** loss of bodily fat reserves.

Blood sugars and electrolytes govern strength, mood, and even consciousness, but a drop in this sort of blood chemistry will be followed again by a rise, as the body compensates for the drop by drawing on—you got it—its reserves in fat and vitamins.

People weak from exertion, collapsed, and on the verge of fainting, can be brought back quickly by eating a high carbohydrate, or sweet food, which restores blood sugars. This is the justification candy bar makers use for their "fast energy" commercials.

Sweating takes water, salts, and potassium out of the body. That sweat is chemical rich, and people do pass out when blood electrolytes drop to the danger point.

Potassium tablets are more necessary than salt tablets for hot weather, since it is the potassium that helps maintain the right electrolyte level.

An all vegetable diet is not terribly bad for you, if you have a reasonable variety of vegetables. The human body actually manufactures protein from carbohydrates, a leftover from our leaf-eating days.

Pack vitamin tablets in on your operations, urban or field, to keep the necessary organ-sustaining vitamins circulating in your system. The body ***consumes*** vitamins daily, and can run out of them. When this happens, you start to develop strange symptoms, like slow healing wounds, loss of night vision, inability to digest some foods, pain in your joints, and skin ulcers or sores.

Read a basic health and nutrition book before embarking on a three-month contract in the Outback. 90 days is about the limit a normal, healthy person can go with a bad diet before health problems make them unfit for duty.

TIME OFF

If you're working a long term contract somewhere, time off should be consideration when details are being discussed. You need to return to civilization to get your perspective on life, have some fun, and spend some of your hard-earned money.

Long months in blast-furnace hot mideastern deserts, cold, isolated mountain outposts, or mosquito infested, sweat soaked jungles put pressure on you, rack your health and twist your outlook on life. Just simple exposure to normal people and good food bounces you back, and makes the next contract a little easier.

The U.S. military in Viet Nam used to give its personnel a choice of R & R cities, with offices and assistance there, provide flights and other transportation, and generally act as a buffer between the farm boy and the foreign city.

As an independent, there are no such services waiting for you.

You have to arrange for your own time off, decide where you want to go, and how to get there. You'll have to pay your own way, and get yourself back.

This is not to mention staying out of jail because of your holiday-making while enjoying your R & R.

In this chapter I'm going to give you the Merc's Guide to R & R cities — and the tips you'll need to do it right. Because there are more places to go than we can fit into this entire book, we'll center on the highspots of the world.

For those of you working in Central America or Mexico, I'm recommending ***Mexico City, Mexico.***

South or Central African-based mercs might be closer to ***Johannesburg, The Republic of South Africa.***

European job holders will find excellent R & R facilities in ***Athens, Greece.***

Bangkok is the southeast Asia's merc choice for a holiday, this Thailand capitol city being one of the old U.S. military haunts for their R & R program. Maybe Bangkok isn't still holding a grudge.

These choices are based on price, attractiveness to the tired but deserving merc, and of course, cultural worthiness. The fact liquor, women, nightclubs, and good airports are available in these cities has nothing to do with it.

We'll begin with Mexico City.

Mexico City is in the state of Mexico, in the country of Mexico. Mexicans don't call it Mexico City . . . they simply say ***Mexico.***

It is not what you might expect of a land imagined by North American (gringos) as full of burros and sumbreros. Mexico City is a massive (New York City scale) ***European***-style city, with an international outlook.

The Mexican ladies are beautiful, friendly to foreigners, and in the city, approachable. Don't try it in the suburbs or beyond. More than a little touch of old fashioned Spanish influence still exists, and the local husbands and fathers are very narrow-minded.

Mexico City if full of elegant hotels (with bars), restaurants on the streets, and liquor is served in the restaurants, just like home.

Although the traffic is wild, the taxi drivers are good, cabs are plentiful, and English is widely spoken by the drivers.

Airport security is normal, and customs baggage checks

are not a real pain. The Mexicans are easy going people, but they are not stupid. It is important to be polite—Spanish is a polite language, if you trouble yourself to learn it—and obey the police and airport officials.

Mexico City has airline connections to almost all Central American countries making it easy to get in and get out. Traveling on established routes and with reliable airlines (Air Mexico and TACA are good examples) makes scheduling yourself back to duty little problem.

The only real travel hitches are in small airports both in Mexico and surrounding countries. You can sit for some time in Belize or Guatemala City as flight after flight is diverted or canceled.

Private citizens in Mexico can own firearms, so be on the lookout for concealed .38's or .357's. The .45 ACP and .308 cartridge are outlawed for private use (military or police only) so if you see someone, even in civilian clothes, who is packing a .45 auto, you know who he is.

The knife is the weapon of choice in Mexico and Central America (if the machete is not available) and no one will take exception if you have one.

Gringo faces in Mexico City are common because of the tourist trade. You can dress and act like a tourist if you want to fade in that way, or buy some casual clothes from a local shop and imitate the hometown boys.

The peso is on the downhill run these days, making prices cheaper for visitors. A dollar sign ($) like we use in the States is also used to denote the peso, so $377 on a menu for a sandwich isn't as steep as it first seems.

You can even use a number of international credit cards in Mexico hotels and restaurants, but only the swanky places.

You'll like Mexico City. It's alive, modern, and has all the sinful elements anyone would want, and is set to serve those who have the cash to pay for a good time.

CAIRO

Cairo, capital of Egypt, is a haven for the wearing on that side of the world. It is a tourist city because of the nearby pyramids and the Nile really does run through town.

What makes Cairo the choice?

If you've been hiring yourself out in the Mideast, you've

learned how two of life's necessities are denied among the desert set . . . liquor and women.

Cairo, Islamic center though it is, has more than enough women and liquor to go around—and you can even get pork there. If you like to have sausage with your eggs in the morning before you go back to some Saudi or Omani shiekdom and do without, you won't have to in Cairo.

There are nightclubs in Cairo. There are belly dancers in Cairo. There is wine in Cairo . . . and there are Egyptian women and girls in modern western clothes that frequent those places with the clubs, dancers and wine.

You'll spend Egyptian pounds in Cairo, broken down by the piastre. There is no currency restriction that really will apply to you if you're changing money for your R & R, and yes, you can get a better rate on the street, so don't change it all at the airport.

Speaking of the airport, Cairo International is a hell of a ways from downtown, but they have a well regulated cab system that keeps you from being overcharged.

Cairo is part of the Mideast, and that means erratic power, water and telephone service. Make your airline reservations ***in advance,*** and don't cut your scheduling too close. I've been in Cairo when I couldn't call across town to the airport, wasn't notified of my first flight being canceled, and got caught in a riot at the airport (customers were fist-fighting police in the lobby at the ticket counter) and have had to bribe airline employees just to be able to get my ***reserved*** seat.

Baksheesh (bribe) is a daily thing in Cairo, and you are advised to be liberal with it for services you might otherwise take for granted. It'll open up doors and grease many skids for you.

Egyptians are poor and sly. Street hawkers prowl with buys in gold, perfumes, money exchange and dope. Watch your wallet—they are all crooks. A street sharp will not hesitate to present himself as a government or police-approved this or that.

Private firearms are not common in Cairo, and cheap knives, iron plumbing pipe and bricks are. Egyptian men and boys are ambushers and sucker-punchers. They rarely take a one-on-one fight.

Security at the airport is typically Mideastern. They'll tear some poor bastard down to his socks and let a hundred others pass by while the agents confiscate their victim's *Playboy* magazines and marked prescription tablets.

Don't look anything but dumb and casual as you go in or out of airport security. Combat-age males are watched close by the AKM-holding guards who expect trouble.

Don't be funny or superior with armed Egyptian military or police. They won't like you because you're suspicious, and they can sure ruin a good R & R with a stay in a stinking cutthroat Cairo jail.

Security extends to the hotels and restaurants in town. Terrorist bombings are a reality and the grenade nets, searches at the door, and metal detectors you'll have to go through just to get into the Sheraton are a necessary pain.

Women are denied to even the Arabic males in the Mideast, the exception being downtown Cairo, ***so do not approach any women who are not obviously part of the bar or nightclub staff.***

This is serious business, sir. You'll need a MAC-10 to get out of the hotspot you'll create if you make a pick-up attempt on someone's sister or daughter.

Cairo can be an adventure all by itself for the merc. The keyword is respect for the women, keep the liquor in the hotel room, bar or restaurant, and stay out of the alleys.

Cairo is the travel center of North Africa. Saudia, PIA, Air Egypt and British Airways serve it to name a few. Europe uses Cairo as a vacation spot, so tourists are tripping over each other in the right places.

European tourists and American tourists dress very much alike, with the more casual appearance going to the Americans.

European men can dress in a fashion considered almost theatrical or effeminate to American men and still be within the bounds of acceptable dress by their own standards. Their jeans will be tighter, their shorts shorter, their shirts more tailored and their jewelry flashier than the American. Understand, you have to deal with the Italians, French or Greeks as tourists in Cairo—or if you want to imitate their dress.

The cowboy craze isn't really that popular in Europe, so if you go in boots and Texas shirts, you might be inconspicuous in Montana, but you are a signal in Cairo, and the message is ***American! American! American!***

A final tip for those of you who plan to travel a lot in the

Mideast; an Israeli entry stamp in your passport will bar you from many Arabic countries and cause suspicion in the rest. Egypt and Israel do have an uneasy peace right now, so you can get into Cairo or Alexandria with Hebrew ink on your passport pages.

The Israelis will give you a removable page for your passport with their visa stamp on it if you ask. Do ask.

JOHANNESBURG

This is not the swinging South African city you might imagine—it's really rather conservative, but it is available and there for the merc doing his contract down that way.

Angola and Zimbabwe, Southwest Africa (Namibia) and Zaire are in the neighborhood, all merc earning grounds now, so the appealing assets of "Jo'berg" (as the locals like to say) is its realiable travel, employment contracts and contacts, and wine, women and song.

South African young ladies are available in either English or Afrikaans-speaking models, and are comparable to American girls of the late Fifties in social outlook.

Afrikaans is a modification of the old Dutch, and is interchangably spoken with English in South Africa. For instance, "In the Samurai tradition," becomes "In die Samurai tradisie," in Afrikaans. If you understand some German or Dutch, it will help.

Jo'berg is pleasant in the way old plantations of colonial cities can be pleasant. Rand is the currency there, and use some caution in your exchange because there are restrictions as to how much you can convert and reconvert when you are leaving.

Pretoria, just North of Jo'berg, is the capital, and both cities have ample breweries, night clubs and bars. The South Africans love to drink—and—fight—so mind your manners with them.

A civilian can own weapons in South Africa, making odds of the bloke you're sizing up for a showdown in packing a piece very good indeed.

South Africa is engaged in a counter-terror war, of course, but you'd never know it from living in Jo'berg.

The war is a very distant thing, despite the apparent desperation the world news attributes to the border patrols and raids out of South Africa to Angola, etc,. Remember, it has not gotten to the point within South Africa that there is an active guerrilla war as there was in Rhodesia (excuse me, Zimbabwe) or Angola.

The business as usual, attitude of the South Africans might be a refusal to face what is upcoming, but it is nonetheless the attitude.

Since English is widely spoken there, you English language mercs will have no trouble, and the foods and ladies will be familiar.

An American stands out like a Martian once he opens his mouth among the British accents, and since the mercenary business is as illegal there as it is in the U.S. of A., you fall into one of three categories; businessman/tourist/merc. Be a tourist. Don't go around telling war stories, because the South Africans don't want any ops being launched from their borders, and they might not buy your vacation in Jo'berg story, although it is true.

On-time, no-nonsense airline schedules go out night and day, with complete international connections. Jo'berg is civilization complete, so you may make phone calls, book leaving times and so on without worrying about anything.

Jo'berg is a comfortable place. Enjoy it.

ATHENS

Greece, on the warmwater downside of Europe, has been a vacation haven for the world. You can blow in from anywhere and melt right in.

Greece is a good choice for those of you who've been pulling Mideastern contracts and want badly to get away from the Arabs. It's also great for European mercs who may have been doing security in Italy, Ireland, Spain, or those unmentionable places along or behind the Iron Curtain.

In the spring and summer, Athens is a paradise for the R & R seeker. International hotels, open bars, swimming, island tours, lots of European women filling those bars and pools, and even cultural tours for the merc with sense of history.

You can see the famous Parthenon (blown up as an ammo dump in an obscure war long ago—that's why it looks so wrecked), the original Olympic tracks and arenas, evidence of the Roman occupation, and the cave where Socrates took hemlock.

Don't bring a firearm into Athens. Guns there are a military only-type thing, and the government is a rather strict, humorless sort . . . a socialist police state, weapons ready.

Currency control is serious as well. Only change what you plan to spend, maybe less, then change a bit more later. You don't really want to leave Greece and have to take an expensive wad of ***Drachmae*** with you as a reminder of your good time.

Greek men are macho, knifewielders, and on the whole are strong and very brave. They might not like your face and you'll have to either make a hasty, scared-tourist retreat out of the bar or make a stand until the police get there, and guess who goes to jail?

Anyway, you came to Athens for the sun and fun, right? You should be paid to fight. Don't give out free samples.

Athens airport is outside town, and many cabbies don't understand English. A lot of motioning and slow pronunciation of your hotel name might get you on the road.

Athens is a city full of poor Greeks, so flashing your cash around isn't advisable. Keep your clothes and wallet low profile on the street. What you do to impress the Air France stewardess at the Lord Byron bar in the Caravel hotel is up to you . . . that's indoors.

Phone service in Athens is sometimes . . . sometimes on, sometimes off. Power outages are frequent in the city. The water is good in the hotels, but don't trust it in local shops. Drink beer instead.

Greece was at one time under the great Islamic empire, so the tradition of protected women is strong there. Use your Mexico and Cairo rule here. Stick to the vacationing European honeys or the local girls who you ***know*** work the bars and hotels.

Athens is a beautiful city, located handily on a most attractive waterfront, with tour ships taking tourists to the islands every day. If you like, book one of the cruises. The women are packed onto those love boats, and a nice cruise might be the unwinder you've been waiting for.

Arrange your in and out airline well. Athens is well conected for international flights, but a bit iffy. It might be Arabic heritage.

BANGKOK

Perhaps you've been to Bangkok courtesy of the U.S. military. We used it as an R & R dump during our bitter war in Indochina, had American troops stationed there, and generally pumped a lot of money and GI's into the economy.

Well, the Americans are gone, and so are most of the services that entertained them. The steets of bars, the sloe-eyed girls, the street hawkers pushing dope, the U.S. MP's . . . all history.

Today, Bangkok has gone back to being Asian, but still has the facilities to take fine care of you.

The airport is clean and modern, serves thousands of tourists a month, and is equipped for the task.

Boht is the currency in Thailand. Currency restriction is no problem for your R & R, and U.S. dollars will spend there, so you might carry a supply of both.

Bangkok is colorful, busy, and a fine taste of the Orient. You'll have all the recreation outlets you can afford. The city is known for its clubs and nightspots, and the slim black-haired Thai girls peddle a brand of welcome hard to leave.

The Thai men are small but not timid, and the strength of some of their laborers along the riverfront is no less than what you'd expect from longshoremen in the States. Most of these guys are as aggressive as they are short of patience in a confrontation with a foreigner. Don't make the mistake of thinking just because your antagonist is 5' 3", he'll be a wimp.

On the other hand, all Orientals are not karate experts either, but they would like to make you think so, affecting threatening stances, and false moves, to throw you off their act.

Thai men will quickly use a knife or broken bottle in a fight, or might just show up later with an army if they think you'll be hard to take one-on-one.

You won't see much in the way of personally owned firearms in Bangkok, even though permission can be secured from the government to own and carry a pistol. The average Thai simply can't afford a gun. A guy there who has one is either an off duty cop, government agent, or criminal. All three types are more trouble than a peace-loving merc on R & R wants.

Thailand is a country on the borders of a war. Vietnam, ***Kampuchea*** (you do remember Cambodia, don't you?), and Laos are in various conditions of conflict everyday. The Thai government is trying to keep the fighting on the far side of their borders, which does not always work.

This is fair warning to keep your distance from border towns. Without proper credentials, you can find yourself arrested, or involved with the hostilities.

The phone system is good in Bangkok, and the airline people are efficient, so civilized arrangements may be made for your departure. This means you can have a reservation and count on it.

English is not commonly spoken in Bangkok, but it isn't unheard of, and the better hotels and restaurants will have multilingual staff... anyway, one of the delightful aspects of a week-long romance with a homegrown sweetie is overcoming the language barrier, right?

Bangkok is a bright spot in Southeast Asia, with the dominoes all fallen as they are. If you happen to be working the Far East, say India, Pakistan, Malaysia, the Phillipines, you really have to consider Bangkok.

General Intelligence Tips:

The easiest way to set up your R & R is to get a travel agency and let them help you. They don't charge for the service, and will be able to make reservations and arrangements you won't even know existed.

Travel agencies are almost everywhere in this old world, from alley offices in Saudi Arabia to air conditioned marvels in the Sudan. Find the one nearest you and use it. It'll save you much time and hassel.

Pack light. Good hotels have laundry facilities. Carry your own alarm clock and notebook, and make sure you have the local phone numbers of the airlines on which you have reservations. Keep your extra money and valuables in the hotel safety deposit if you can, don't leave it in your room for the maid or your girl friend of the evening to abscond with.

If you're traveling with an expensive camera, try and have your sales slip on it, as customs people are watchful for taxable items. This applies to watches, stereo equipment or anything of value which most countries have tariffs on.

Keep your passport very close. It should always be with you, as you might be asked to produce ID on the spot if the police or military have roadblocks you must pass.

Don't stand out any more than you must. Staying hidden in the crowd is the best survival. Smile, be nice to people, and shy away from trouble. You might defend your pride in a fistfight, but the loss of your R & R, and possibly your job is not worth it.

Of course there are many other places that would make good R & R spots. Those of you who have been out there in the world are probably brimming with suggestions right now. This is only a highlight guide.

All work and no play makes the merc a dull boy. Take some R & R if it's offered, and take care.

GETTING PAID

Money, the point of the entire exercise in the first place, carries with it a set of problems not readily apparent to those innocents who think you simply get your paycheck and bank or spend it.

At home, a dollar was a dollar and a dime was a dime. Down in the South Outback, the local currency might look like Monopoly money and change value once a week. A reassuring factor of any merc job is discovering you're going to be paid in local money, and then discovering said money is not on the International Currency Exchange.

Taxes on that money come next. The host government will want to tax you (given half a chance) and the U.S. government will tax you given no chance at all. If you slip and slide around, trying to funnel funds back into the States covertly, you'll quickly discover one of our most alert branches of government is the Internal Revenue Service, and they monitor anything of value coming or going past America's borders.

Currenty transfer restrictions are part and parcel of U.S. practice (and other countries) so check and see if you can move that fortune you plan to make out of the country you made it in.

Here's what you need to know to deal with foreign pay.

International Currency Exchange

Major countries keep their currency listed on this Exchange, which is like a stock market, giving daily rates of value for the different currencies against each other.

Most international newspapers publish a daily rate of the major currencies, so you can see if the Peso's you're being paid in are gaining or losing compared to the currency you want to convert them into.

Many currencies are not listed on the money market, and have an artificial value set by the nation that issues the money. Inside the nation, you can get the "official" exchange rate, but on the black market, or outside the country, the money will be worth far less.

A currency regulated like this indicates bad economic problems. Usually, strong restrictions on how much foreign currency that can be taken ***out*** of the country go hand in hand with this situation, so as to keep as much foreign money in country as possible.

Exchange Rate

The rate of exchange is the value of one currency as compared to another, usually changing on a daily basis.

The rate of exchange is something you have to watch closely, because you can make or lose money on it. By example, let's say you're being paid in Saudi Rivals, which is exchanging at the rate of 3.5 to the American Dollar. That means a Dollar is worth 3.5 Rivals.

$1,000 Dollars at that rate would be equal to 3,500 Rivals.

But, on the next day, the rate changes, becoming 3.2 Rivals to the Dollar. Now your same $1,000 Dollars are worth 3,200 Rivals. If you're changing Dollars to Rivals, that's bad, but if you're changing Rivals to Dollars, suddenly you are making money.

The 3,500 Rivals at 3.2 become $1,120 Dollars.

The sword cuts both ways.

Always carry a small calculator and use these formulas.

Dollar multiplied by Rate equals Local Currency Value.
Local Currency divided by Rate equals Dollars.

Currency Restrictions

This is when a country regulates how much money can be transferred in or out of its borders. Restrictions exist for two reasons, taxation or bad economic conditions, where the idea is to stop outgoing foreign currencies.

You may find the situation of banks and hotels not even accepting their own currency, and demanding foreign money from you because you are a foreigner.

If the reason for restriction is taxes, such as the United States, you will be ordered to declare to customs the amount of currency you are bringing into the country.

Ask about restrictions before you change your foreign currency into the local tender, because there may be a limit on how much you can convert back as you leave. Only change enough each day or week to get by.

In economic emergencies, bank transfers of money out of the country in any amount may be regulated or forbidden, to keep residents from liquidating their bank accounts and moving them to another country, slowing the drain of funds.

Bank Transfers

The best means of getting money out of a country quickly is to make a bank transfer.

To do this, go to a bank in the country where you're working, and ask to make a bank transfer. You will be given a form to fill out indicating the amount of the transfer, and the bank and account number in the country where you want the money to go.

The bank officer will then take your cash, convert it into the currency of the country where it is headed (if you request a conversion), using the current exchange rate, and give you a receipt.

Don't lose the receipt. It has the transfer number on it, in case of problems and the transaction has to be traced.

There is a small bank charge for this service, but it is reliable. Expect the transfer to take several days, because the money will vanish into the electronic limbo of all wired currency transactions, keeping a standing pool of many millions out there on which the bank can use your money and not have to pay interest.

Money coming into the United States like this is automatically reported to the Internal Revenue Service, so expect Uncle Sam to know about it.

Swiss Accounts

These are the famous "numbered" accounts. The truth is, the Swiss banks are not the only banks on earth to offer numbered accounts, and they do know who you are, and will reveal your name to the police if given adequate evidence you are breaking the law somewhere and using the account to hide money.

Some types of Swiss accounts pay interest. Some don't. Most require a hefty minimum deposit to open an account.

The only advantage to a Swiss account is better security for you than most other banks. They will tell you if a law enforcement or tax agency is inquiring about your money, but again, they will open their books to the proper authorities if shown enough evidence they should. The Swiss bankers are not partners in crime with any account holders.

Taxes

Working overseas does not absolve you of taxation by both the country in which you are working and your own country.

Better check and see if the land of opportunity where you're going want to tax out a big chunk of your salary, because your hometown tax gang will want a piece too.

If you're being paid in cash for your labors, expect to be able to show your home country tax agent some proof of your income, because they will press you hard for it come audit time.

Where most people get caught in the tax trap is transferring money home, or in carrying large amounts of currency or gold home and being caught with it in Customs.

GOLD

Buying gold is not for amatuers, and a study of the market and values are necessary before you invest.

If you're going to invest in gold, do it in coins or bars, so the value can be calculated quickly. A beautiful watch or bracelet is not as easy to calculate into cash as plainly marked 1-Ounce Kruggerrand.

The value of gold changes daily, and it is quoted by the ounce, and if you have gold, you'll want to keep track of its worth in the newspapers, because that's what you'll be able to get for it when you sell.

You can make or lose money by investing in gold, as the gold market values rise and fall, so try to buy your gold when the prices are down.

The Karat rating of gold is an indication of its purity. Pure gold, often called solid gold, is 24-karat. The rating of 14-karat therefore, means the object is 50% real gold, the rest being alloy.

Jewelry has to be an alloyed gold in order to be hard enough to withstand wear, because pure gold is soft, one of the reasons a bracelets value, being part gold and part alloy, is hard to quickly calculate by weight.

Black Market

In most countries the Black Market is illegal, and getting caught trading with street money changers can get you jailed.

Black Market money changers give better rates than the official exchange. They can afford to pay you a higher rate than is posted by the local government or even the International Currency Exchange, because they have a buyer for your money that will pay them more.

Your money is being sold by the Black Market changers to drug dealers, rebels, and revolutionaries, who need it for foreign purchases.

Naturally, Black Market changers only want certain currencies, usually U.S. Dollars, and the tighter the restrictions and the more severe the economic conditions where you are, the better the Black Market rate is likely to be.

You'll run into Black Market changers on the street near tourist center hotels, and at airports and banks, anyplace money changing occurs.

Be careful in doing business with them, but done right, you can get more of the local currency in a swap.